MULTI-PURPOSE TOOLBARS

Some tools for agriculture

Introduced by Dave Gibbon

Intermediate Technology Publications 1987

Practical Action Publishing Ltd
27a Albert Street, Rugby, CV21 2SG, Warwickshire, UK
www.practicalactionpublishing.org

First published 1987
Digitised 2013
Printed on Demand

ISBN 10: 9 46688 78 8
ISBN 13: 9780946688784
ISBN Library Ebook: 9781780442327
Book DOI: http://dx.doi.org/10.3362/9781780442327

A catalogue record for this book is available from the British Library.

Since 1974, Practical Action Publishing has published and disseminated books and information in support of international development work throughout the world. Practical Action Publishing is a trading name of Practical Action Publishing Ltd (Company Reg. No. 1159018), the wholly owned publishing company of Practical Action. Practical Action Publishing trades only in support of its parent charity objectives and any profits are covenanted back to Practical Action (Charity Reg. No. 247257, Group VAT Registration No. 880 9924 76).

Intermediate Technology Publications,
9 King Street, London WC2E 8HW, UK

ISBN 0 946688 73 7

ACKNOWLEDGEMENTS

This book, and its parent volume, *Tools for Agriculture*, was assembled by a team led by Patrick Mulvany, Agriculture Officer of ITDG. The team drew on help and advice from a large number of individuals and organizations. It is not possible to thank them all here, but particular mention must be made of the Overseas Development Administration of the British Government, the German Approriate Technology Exchange (GTZ/GATE), the Swedish International Development Authority (SIDA), and Barclay's International Development Fund.

Printed in England by Russell Press Limited
45 Gamble Street, Forest Road West, Nottingham NG7 5GY
Tel: 0602 784505

Contents

Page

HOW TO USE THIS GUIDE

This is one of the twelve sections of the latest edition of *Tools for Agriculture*. Each section is introduced by a specialist who sets the range of tools, implements and machinery available against the background of good farming practice, and the factors best considered when making a choice.

It is intended the guide should be used by the following categories of people:

- Farmers' representatives who purchase equipment on behalf of their clients;
- Advisers who seek to assist farmers and farmers' organizations with the purchase of equipment;
- Development Agency personnel who need to purchase equipment on behalf of farmers and farmers' organizations;
- Prospective manufacturers or manufacturers' agents who wish to have information on the range of equipment currently available.
- Development workers, students and others who wish to learn about the types of equipment available.

We expect the reader to use the guide in one of the following ways:

- to find the name and address of the manufacturer of a specific piece of equipment whose generic name is known e.g. a treadle-operated rice thresher or an animal-drawn turnwrest plough.
- to find the name and address of the manufacturer of a piece of equipment whose general purpose is known e.g. a machine for threshing rice or soil tillage equipment.
- to find out about specific types of equipment or equipment used for specific purposes.
- to find out about the range of equipment available from specific manufacturers or manufacturers in particular countries.
- to learn more about equipment used for the different aspects of crop and livestock production and processing.

Guidance across the broad range of small-scale farming equipment is available, with indexes and cross references, in the parent volume.

Within each section the information is presented in three ways:

- A clear *introduction* which lays out the most important points to bear in mind when purchasing a particular type of equipment. (The emphases vary from section to section — showing the difficulty of decision-making when selecting equipment for smallholder agriculture.)
- *Comprehensive Tables* which list the manufacturers of certain types of equipment and give some further information about specific items, or the range of items manufactured. In many tables it was impossible to give the full address of the manufacturer and the reader is referred for these to the Manufacturer's Index.
- Pages laid out in grid pattern in which the compiler has attempted to present the equipment in a logical order, that in which operations are carried out, and within each type of operation the progress is from hand-operated, through animal-drawn, to motorized equipment. Sometimes one particular type of equipment is illustrated to represent a group — many of which may differ in detail, though not in use. Wherever possible the trade name of the equipment is used, in order to facilitate enquiries to the manufacturers.

Having located a manufacturer for the type of equipment in which you are interested, we suggest you write direct to the manufacturer for further details: current prices, availability, delivery times and so on. (Remember that, where known, telephone and telex numbers have been included in the manufacturers' index).

Every attempt has been made to ensure accuracy of the details presented in this guide, but doubtless changes will have occurred about which the compilers are unaware. We apologize to any reader to whom we may have given a false lead. A note will be made of up-to-date information which becomes available to ITDG.

It must be stressed that this guide relies on information supplied by the manufacturers and that inclusion of an item is no guarantee of performance. Whilst every care has been taken to ensure the accuracy of the data in this guide, the publishers and compilers cannot accept responsibility for any errors which may have occured. In this connection it should be noted that specifications are subject to change without notice and should be confirmed when making enquiries and placing orders with suppliers.

GENERAL INTRODUCTION

Problems of farmers in developing countries

The main economic characteristic of agriculture in developing countries is the low level of productivity compared with what is technically possible. It has been shown in many and varied circumstances that although farmers may be rational and intelligent, technological stagnation or slow improvements can still be the norm. This contradiction can be explained by understanding several unusual, troublesome features of agriculture. First, because agriculture is basically a biological process, it is subject to the various unique risks of weather, pests and disease which can affect the product supply in an unpredictable fashion. Despite exceptional biological risks, most farmers nowadays rely to various extents upon cash derived from sales of produce. But agricultural products have consumer demand patterns which can turn even good production years — when biological constraints are conquered — into glut years and therefore financial disasters. The biological nature of production also results in a large time-gap, often months or even years, between the expenditure of effort or cash and the returns. Once cash inputs are used, an unusually high proportion of working capital is required, compared with industry. The final problems created by the biological nature of production lie in the marked seasonality. The peaks of labour input create management problems, and perishable commodities are produced intermittently; both create additional financial and technical storage problems.

A second characteristic of agriculture is from the small scale of most farming operations, often coupled with a low standard of education of the operators, which gives farmers little economic power as individuals and little aptitude to seek such remedial measures as do exist. There are many examples of appropriate technology but small farmers will often need intermediaries, such as extension workers and project personnel, to open their eyes to the potentialities. Given the vulnerability of small farmers to biological and economic risks, those intermediaries have special responsibility to assess the impact of any new technology for each particular set of local circumstances.

A third factor which affects efficiency in agriculture is a political one. It is in some ways ironic that in countries with very large numbers of small farmers, producers tend to command little political power despite their combined voting strength. Indeed they are often seen as the group to be directly and indirectly taxed to support other, generally urban-based, state activities. As a contrast, in rich countries, we often see minorities of farmers with little voting power receiving massive state subsidies, much of which supports technological advancement. The rationale of farmers referred to above thus leads to the exploited, small farmers producing well below potential and the rich, large-scale farmers producing food mountains that can only be sold at further subsidized prices.

The 1970s food crisis, the recent failure of agriculture to match rising food demands in many countries, particularly in sub-Sahara Africa, and the failure of industry to fulfil its promise of creating employment and wealth has turned the attention of policy-makers back to the long neglected and often despised agriculture sector. New technology for the large number of low-income, small-scale, poorly educated farmers will be necessary if agriculture's enhanced role is to be successful.

What are the technology options?

Innovation and technology change has been and will be the main engine of agricultural development. Technology

Innovative equipment can be simple in construction: a four-furrow row seeder.

change can be described as the growth of 'know how' (and research as 'know why'). But technology is not just a system of knowledge which can be applied to various elements of agricultural or other production to improve levels or efficiency of output. Technology application requires and uses new inputs. In contrast, technique improvement is the more difficult art of improving production essentially with existing resources. Pity the poor agricultural extension worker sent out to advise experienced farmers with no new technology, but only improvements in technique to demonstrate!

It is possible to exaggerate the lack of prospect for improvement and the consequent need for new investment in farm resource use. Changes both on and off-farm are influencing the economies of traditional systems. For example, with farm size halving every twenty years or so in some regions — as a result of population growth, and with increasing demand for cash from farming activities for production items such as seed (which used to be farm produced) or for consumption items such as radio batteries and so forth, there are new challenges to the traditional rationale and the old system optima. But despite the need to adjust the existing resources to find new optima, the opportunities for really big gains will undoubtedly come from new technology which will often require radically different ways of doing things. A change in the resource base or the injection of a new piece of technology into an interdependent agricultural system may alter various other constraints and opportunities within the overall farm system. One function of a reference book such as this is to act as an encyclopaedia illustrating alternative ways of coping with new challenges. Readers do not have to reinvent the wheel each time a new transport system emerges. Self-reliance has little merit over technology transfer when it comes to solving food availability problems in a rapidly changing world.

This book displays a very wide range of technology and describes both what the technology can achieve, and how and where most information can be discovered. It shows that there is already in existence a mass of tested technology for small-scale farmers. The farm technology itself is laden with opportunities for improving the returns to land, water, labour and other crucial resources. The careful farmer, with help, can have many options.

The role of information

In the theory of classical economics, information on the contents of the technology itself is assumed to be a free good, readily available to all. This is clearly absurd in any industry, but particularly so in agriculture. One of the main justifications for public support of agricultural research and extension, in developed and developing countries alike, is the inability of farmers to search and experiment efficiently and thus to find out what technology is available.

However, knowledge of the existence of appropriate technology will not be sufficient to ensure adoption. Attitudes toward it may need to change, the hardware has to be physically available and those convinced of its value need financial resources to acquire it. One good example of this is family planning technology, where knowledge has generally outrun the capacity of the delivery systems. Similarly, local testing of the appropriateness of various items is very desirable. This in turn, will require more local agricultural research stations to accept responsibility for adaptive research and technology testing. Nevertheless, knowledge is obviously a necessary prerequisite to adoption, and publications such as this have an important part to play in information dissemination.

The impact of technology

Selection of technology for inclusion in this book does not imply endorsement of a particular product. Indeed, supporters of the appropriate technology concept often have an ambivalent attitude to new technology. New technology always changes the system and in particular it is likely to change who benefits from it. Appropriate technology advocates believe the kind of cheap, simple, small-scale, locally produced, reliable or at least mendable technology will increase incomes and improve or at least avoid worsening income distribution. This is possible, but it is still hard to prove that any technology has the ideal intrinsic qualities that will somehow create wealth and at the same time favour the poorest groups in society. On the contrary, experience shows that the income-distribution consequences of change are generally unpredictable. Since new technology normally requires access to resources, it generally favours the better off; the mode of use of technology, and thus its impact, is not a readily visible quality.

To reject all modern technology on grounds related to fears about income distribution is to argue like the elderly man who said that 'if God had meant us to fly He would not have given us railways'. Societies must accept the benefits of new technology and devise means to reduce the social costs associated with any worsening of income distribution — the greater the gain in aggregate income from innovation, the easier this should be to achieve. We are aware that the direct users of the book will seldom be the small farmer client that the contributors and compilers generally have in mind in selecting equipment. But those who have access to this book, such as extension officers, government officials,

Well-designed hand tools can reduce drudgery: harvesting in Morocco.

teachers and local leaders must give guidance with care and with wisdom.

How to choose

In selecting new technology, either for testing or promotion, numerous criteria can be devised to aid judgement. These will include the degree of technical effectiveness, financial profitability, the economic and social returns, health and safety factors, the administrative and legal compatibility with existing conditions. The criteria will not necessarily be independent or even compatible. A financially profitable piece of technology may depend upon underpriced foreign exchange or tax allowances and be economically unattractive. It may substitute capital expenditure on machinery for labour and be socially unattractive. A particular criterion such as technical efficiency, may have several elements to aid judgement — such as the technology's simplicity and labour-intensity, its ecological appropriateness, its scale and flexibility, its complementarity with existing technology and so forth. These elements are not inherently equal and in some circumstances one will be regarded as carrying most weight, in other circumstances another. Choice of technology is a matter of judgement and all the modern aids for technology assessment, for cost-benefit analysis and the like cannot hide this fact. Analysis is an aid to and not a substitute for judgement; the social consequences — which are agonising — must be weighed against the various real benefits that are apparent.

The technologies presented in this book reflect the belief that whilst all technology will alter the economic status of large numbers of people (often in the direction of greater inequality of income, greater commercialization, more wage labour and increasing landlessness) some technologies are more likely to do so than others. You will find few tractors or combine harvesters in this book, for example, but great emphasis on, for example, animal-drawn tool-bars and powered threshers. Technology varies in its degree of reach-down to the low-income farming groups who, if they are not the main target of rural development, are from our viewpoint a key component. The cost of lost output through using less efficient equipment — hand pumps rather than tube-wells, resistant seed rather than crop protection, hand tools rather than tractors, small livestock rather than cattle and buffaloes — is small. Indeed, the productivity of labour-intensive gardening and allotments can often exceed that of modern capital-intensive farming systems — as was shown in Britain during and after the Second World War. Whilst situations do occur where demand for increased food supplies force governments to chase home-produced food without too much thought about the social impact of the production system, such dire circumstances are rare. They might occur where the bulk of low-income people are food purchasers — urban dwellers and landless rural labourers, and in these cases large-scale, capital-intensive state or private farming with the most modern technology, might be justified. But it is only rarely that the trade-off between technical and economic efficiency and equity criteria is painful. Research in many countries has shown that modernized peasant-based systems are generally equally or more efficient and to most views more equitable, and thus it is the small farmers who are seen as the main beneficiaries of *Tools for Agriculture* — even if they are unlikely themselves to be the main readers of this book.

Feedback

Whilst there are a number of people who know and understand the hardware described in this book, there is less understanding of the ways in which technologies are 'delivered', or options presented to the small farmers themselves. ITDG is therefore always pleased to have critical and appreciative feedback — from the aid agencies, extension workers, credit agencies, schoolteachers, businessmen, politicians and others who use this text, on the content and format, equipment that is missing, new problems, the effectiveness of the equipment, the service of the manufacturers, and new ideas for delivery. The hardware available grows rapidly in diversity and power, but, just like computers, it will be useless without the software support. In the case of agriculture, technology software stems from the efforts of interested individuals and groups who are close to the small farmers. We look forward to hearing from you!

Ian Carruthers
Wye College, University of London

STEPHEN BIGGS

Modern technology is easily applied, if one has the resources: fertilizing maize.

MULTI-PURPOSE TOOLBARS

NIAE

Multi-purpose toolbars have evolved relatively recently in response to changing circumstances in many developing countries.

With the very oldest of animal-drawn implements, the single tool was used as a multi-purpose implement — for primary land tillage, seed-bed preparation, seed covering and even inter-row weed control. Tools of this kind still exist in parts of the world (see Section 1) and are used in much the same way they were when they were first developed. They have the advantages of cheapness, versatility and ease of manoeuverability. They do suffer from some limitations in the efficiency with which they can carry out certain critical operations.

More recently, in Asia, appropriate implements have evolved to suit particular soil and climatic conditions, and although equipment for rice paddy preparation is often adequate, the range of animal-drawn tools for rain-fed, dryland agriculture is of limited versatility.

Attempts to modernize agricultural production systems in the developing countries have either focused on the transfer of western capital-intensive technologies or on the transformation of indigenous technologies through the introduction of adapted components from exotic systems. This has included new crops, varieties, fertilizer technology, crop protection techniques and crop processing technology. Until about 50 years ago there was little serious research and development work on improved tools, either hand- or animal-powered. It was assumed that either very simple adaptations to existing tools or the importation of European tools, would be adequate.

When it was finally realized that Western mechanization technologies would not transfer directly, some work began on improved tools. Much of this work focused on improved ploughs, but a number of engineers, working independently, developed the

Table 1. Working times for exclusively manual operations.

Operation	Groundnut			Maize	Millet		Cotton	
	Ivory Coast	Senegal	Chad	Ivory Coast	Senegal	Benin	Ivory Coast	Cameroon
	days/ha							
Tillage	40	15-20	10	40	—	13	8	15-20
Ridging	—	—	—	—	—	—	20	—
Fertilization	10	—	—	2	—	—	5	—
Sowing	5	10	9	6	4	7	8	5-10
Resowing	—	—	1	1	—	—	2	—
Weeding	30	15	20	24	12	31	60	30
Thinning	—	—	—	5	2	—	5	3-4
Harvesting	25	19	—	6	16	29	50	40-50

Note: Except where listed the source of tables is Binswanger, H.P., Ghodake, R.D., and Thiestein, G.E. (1979), 'Observations on the economics of tractors, bullocks and wheeled tool-carriers in the semi-arid tropics of India', in *Socio-economic Constraints to Development of Semi-Arid Tropical Agriculture*, ICRISAT, India, Feb. 1979.

concept of a multi-purpose toolcarrier, or toolbar. The principle of this idea is that the equipment consists of a main frame on which a range of soil-engaging implements can be attached to carry out the production tasks.

Multi-purpose toolbars can be divided into two main groups:

1. Simple, 'single row' equipment based on relatively modest modifications to the single furrow plough frame or light-weight 'T' or 'A' shaped frame (see pages 50, 51, 52 and 53). There are many variants of these and they have been developed for a wide range of conditions and environments. The main advantages of this group are cheapness and ease of handling and maintenance. They are usually suited to primary and secondary tillage operations and to simple, inter-row weeding work. There may be some limitations in the quality of work achieved, particularly in relation to planting and inter-row hoeing which may be critical operations in some environments. These implements may be drawn by single animals with neck or head yokes connected with chains and a simple spreader, or two animals with a single chain between them attached to a pole neck or head yoke.

2. A broader, two wheeled (or skids) implement that bears some resemblance to a tractor-mounted toolbar (see pages 54 and 55). This kind of implement is well suited to precision strip tillage or bed systems of soil, water and crop management and where time and sweep cultivation is feasible. Since the implement is often wheeled, it can also be converted to a cart, although most experience so far would indicate that the equipment is either used as a cultivator or a cart. Such equipment is drawn by two or more animals, though for light work, it would be possible for only one animal to be used.

Hitching systems vary depending on whether animals have a head or neck yoke, or whether collars are used. If a poke yoke is used (the most common arrangement) the dissel boom is usually attached to the centre of the yoke providing stability for the attached toolbar, and permits flexibility in adjusting working pitch of ground-engaging parts (see page 48). If collars are used on the animals, a more complex hitching arrangement is necessary, involving spacers behind each animal and a swingle tree attached to the horizontally held boom.

A number of engineers have also developed a range of intermediate types of implement, and manufacturers now offer a complete range of implements from very simple cultivators to very sophisticated wheeled toolbars which are capable of carrying out all the operations that can be done by tractor-mounted tools.

Some of the pioneering work was done on these types of implement over 50 years ago and recent work indicates that, in an appropriate crop system, these implements may have real advantages. Farmers are still reluctant, however, to adopt them on a wide scale: some of the reasons for this will be explored below.

The major tasks in crop production systems are: primary land tillage, secondary tillage (where necessary), fertilizer or manure application, seed-bed preparation, planting, post-emergence weed control, crop protection, harvesting, transport, processing and storage. In most small farm production systems all these operations are carried out by hand methods. However, where draught animals are available it is possible to improve the efficiency and timing of certain operations, reduce very onerous tasks and increase the area cultivated through the use of appropriate implements. The use of a multi-purpose toolbar enables farmers to carry out many of these operations and use their own labour more effectively.

Toolbars are considered to be particularly appropriate wherever tractors give a low rate of return (over most of the developing world), where hand methods do not result in adequate returns to provide basic family needs (probably over much of the semi-arid tropics), and where there is an inadequate range of single-function implements.

Advantages

It is necessary to stress that many of the benefits of this kind of equipment are only likely to occur when a number of complementary conditions are present. These include the presence of suitably fed and trained animals, competent handlers and machinery maintenance skills and the potential for additional returns from the use of this equipment above those received from present technology.

The possible benefits are:

- Versatility: the possibility of carrying out a range of tasks using the same basic tool with minor additions or adjustments.
- More effective soil and crop management technology, particularly in relation to seed-bed preparation and post-emergence weed control.

Table 2. Number of working days required to complete a working pattern (planting, weeding, thinning) in manual and animal-draught cultivation.

Crops	Pure manual cultivation (required days of manual work/ha)	Advanced animal-draught cultivation (additional days of manual work/ha)
Groundnut	25-35	10-14
Maize	36	10-14
Millet	18-38	18-24
Cotton	38-75	24-32

- Ease of operation. Once the equipment is adjusted, operations are generally easier than with conventional equipment.
- Lower cost. The total cost of the implement plus attachments is usually less than the cost of the equivalent range of single-operation implements.
- With the more elaborate toolbars, it is possible to cultivate a larger area of ground than with single-row equipment.
- Toolbar-based and other animal-powered technology is potentially labour enhancing rather than displacing as it can generate additional work in maintenance and also increase the opportunity for more frequent operations so that the quality of work is improved. (See Tables 1 & 2)
- Minimum tillage and other conservation techniques are possible using toolbars.
- Recent work in India indicates potential yield benefits on certain soil types. See Table 3.

Alternatives

The purchase of a multi-purpose toolbar may represent a considerable investment for many farmers and it may be beyond the capacity of poorer farmers acting independently. However, it could be possible for groups of farmers to purchase or hire such equipment, and this strategy may make sense where plots are very small and purchasing power is low.

Many farmers have access to very few alternative strategies in their current situation. The range of available equipment is low and of low quality, and the severely limited output potential of hand methods, particularly in drier rain-fed areas, results in a less than adequate output to meet basic needs.

Many field workers have attempted to develop local, low-cost versions of toolbars making maximum use of local skills and materials. While this may be feasible for small numbers of tools, it can rarely be sustained to provide the needs of the mass of the farm population. Quality control is also a major problem with production systems that start up with little experience. This does not rule out the possibility of local manufacture as, in fact, a number of the main manufacturers listed below developed in order to create such industries and, in most cases, they borrowed ideas from other sources.

Other alternatives are the hire of additional labour or the hire of tractors and appropriate equipment. For the bulk of the world's poor farmers these are not realistic alternatives, either because of a lack of sufficient resources at the appropriate time, or of an absolute lack of resources.

Choosing your equipment

Costs and benefits

There are many different toolbars to choose from (if they are available in a particular country). Major manufacturers produce a very wide range of equipment to suit many different situations. Inevitably choice is influenced by the cost of the implement and the expected return from its use in the production system.

The following represents approximate relative costs of toolbar equipment, and alternatives:

Table 4.

Equipment	Capital	Running
Single row or basic toolframe	100	10
Full range of attachments	100	10
Double or multiple row toolbar	500	40
Range of attachments	300	30
Single operation implements (per item)	80-100	10

This return may be influenced by a number of factors outside the control of the farmer, such as soil type, fertility, climate, topography etc, but it may also be influenced by the quality of operations — land

Table 3. Yield increases attributable to cultivation practices with wheeled tool carrier at ICRISAT Centre.[a]

	Steps in Improved Technology (SIIT) experiments, with improved soil management[b]				Watersheds	
Variety:	Local	Local	Improved	Improved	Improved	Improved
Fertilizer:	Local	Improved	Local	Improved	Improved	Improved
					Maize followed by chickpea	Maize and chickpea intercropped
Alfisols	234	802	262	1285		
Medium-deep Vertisols					−190	−206
Deep Vertisols	546	533	479	1046	+166	+625

Source: Ryan, Sarin, and Pereira (1979, Appendix Tables 1 and 2 and Table 3). A fuller description of the experiment is given there. See also Annual Report Farming Systems Program, 1976-7.

a. Averages are for 1976-7 and 1977-8 and are expressed in rupees per hectare of net cropped area. The experiments reported here were carried out with seeding attachments that were more expensive than the ones used for the calculations in Table 4. However, the engineers expect to achieve the same precision with the cheaper devices budgeted.

b. During the year 1976-7 the improved management treatment also included a minor level of insect control, but the differences in 1977-8 are fully attributable to soil management techniques.

Table 5. A comparison of annual bullock labour inputs for some operations by traditional methods and with the wheeled tool carrier.

Soil type (village)	Operation	Existing village practice (average of 1975-6 and 1976-7) (pair hr/NCA)	Traditional method as defined on ICRISAT research watersheds 1977-8 (pair hr/NCA)[a]	Broadbed-and-furrow system with wheeled tool carrier 1977-8 (pair hr/NCA)
Medium-deep vertisols (Kanzara)	Preparatory tillage	45.7	70.3	25.8[b] 18.0[c]
	Manuring & fertilization	2.4	—	7.1[b] 5.1[c]
	Sowing, transplanting etc.	12.5	6.3	7.4[b] 4.5[c]
	Interculture	24.8	—	6.1[b] 5.2[c]
Deep vertisols (Kalman)	Preparatory tillage	25.6	30.5	24.3[b] 18.4[c]
	Manuring & fertilization	1.5	—	6.6[b] 3.3[c]
	Sowing, transplanting, etc.	10.3	12.2	6.6[b] 3.2[c]
	Interculture	7.6	—	7.6[b] 7.3[c]

a. NCA — Net cultivated area. b. Sole maize followed by chickpea. c. Maize intercropped with pigeonpea.

preparation, timing of planting, timing and accuracy of weeding etc. Though farmer management ability will influence this quality, it is also affected by the quality and versatility of the toolbar. In general, the simple toolbars will show little benefit in quality of operation over conventional equipment, whereas the larger toolbars (provided that complementary technologies are introduced) can show considerable improvements in quality of operation.

Unfortunately there has been very little reliable work carried out on the economic evaluation of multi-purpose toolbars, but work in West Africa, India and Botswana has given some indication of the potential of this equipment. The Indian work indicates that the benefits are only fully realized when the equipment is incorporated into a well-planned soil- and crop-management system.

The main economic benefits would appear to accrue from better timing of operations, better weed control and better soil and water management. Labour productivity is generally increased and there is scope for greater employment.

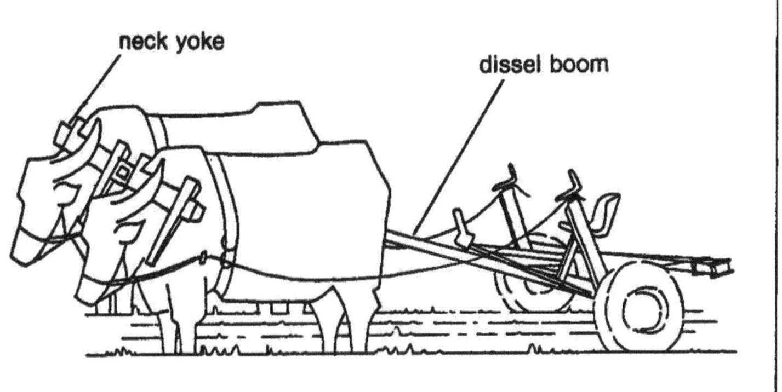

Pair of animals with neck yoke and dissel boom attached between them and to multi-purpose wheeled tool carrier.

Social impact

There is little evidence so far on the social impact of multi-purpose toolbars, as the numbers produced are still relatively low, apart from West Africa, (see Tables 6 and 7) and it is too early to assess these effects.

Some of the potential benefits are likely to be similar to those produced by the introduction of any new cultivation equipment — namely, the removal of drudgery and the opportunity for more members of the household to assist in the crop production processes. It is thought unlikely that the technology would be labour displacing, though this is possible where plots are small and such implements are controlled by a few wealthy farmers. Only through the ownership and control of the means of production will small-holder farmers benefit from these toolbars.

The larger toolbars allow the introduction of major changes in land and crop management techniques, such as land forming and double or multiple row planting and weed control. The simpler range of toolbars can also be used in ridge systems (for instance) but their speed of

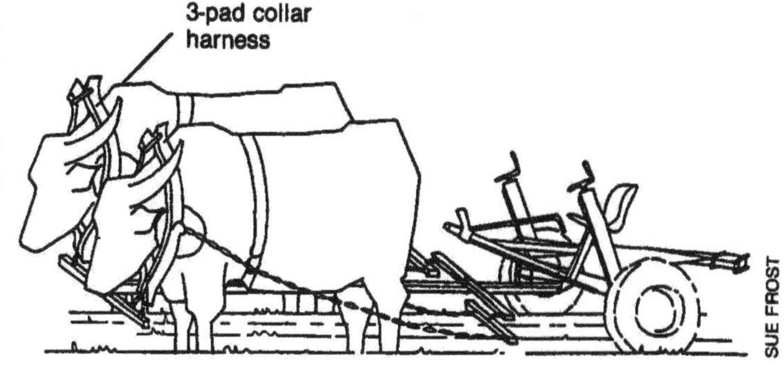

Pair of animals with three-pad collars hitched to multi-purpose tool carrier with horizontally held boom, spacers and swingle tree.

Table 6. Equipment numbers in West Africa

Items	Upper Volta	Mali	Mauritania	Niger	Senegal	Chad	Benin
Bullock-drawn ploughs	12,050						
Donkey-drawn ploughs	4,470						
Total ploughs	16,520	106,700	2,400[1]	3,000	8,000	58,056	7,450
Tool carriers (multiculteurs)	2,500	40,555		4,300 }	204,000	1,727	185
Cultivators (donkey, horse-drawn)	21,000	14,058		900 }			
Harrow		10,739					
Seeders		9,707	100[1]	900	220,000		
Carts		52,204		3,300	89,600	14,606	
Ridgers (ridgers and/or weeders)				1,500	5,000	(3,883)	
Lifters				3,300	88,000		

1. Tentative figures.

Table 7. Average density of animal-drawn equipment in West Africa (ha/unit).

Items	Upper Volta	Mali	Niger	Senegal	Chad
Plough	190	12	900	278	19
Multiculteur or hoe	160	31	690	11	199
Ridger	—	—	—	445	199
Harrow	—	116	—	—	—
Seeder	8,800	129	3,020	10	—
Lifter	—	—	820	25	—
Cart	325	24	820	25	76

operation is inevitably slower.

Other considerations are: the presence of adequate repair and servicing skills and facilities, either on farm or in local villages; the presence of skills in animal draught training and management; and the role of livestock in the current farming system.

The significant technical characteristics that need to be borne in mind are the need to obtain an implement and fittings that are appropriate to the present and future needs of the farming system. The operation and maintenance of this equipment generally requires a higher degree of skill than with most older equipment. A minimum range of repair and adjustment tools are necessary, as is the commitment of a certain amount of time to ensure efficient operation.

It is also usually essential to develop concurrent improvements to animal linkage and harnessing systems to ensure efficient operation. Maximum benefit can be obtained with healthy, fit, well-trained animals and with equipment in good condition.

The type of animals that are available is also an important factor. Small oxen or donkeys, often found in the drier and poorer areas, will only be able to pull the smaller types of toolbar. Camels, larger oxen, mules or horses will be more appropriate for the larger equipment.

Special considerations

Maintenance is obviously important, as badly adjusted equipment will not perform any better than alternative single-operation equipment. A range of spares or repair facilities are essential for uninterrupted operation in the growing season.

The equipment alone will not lead to significant benefits without the development of an understanding of its potential value in a cropping and farming system. Appropriate animals, training and harnessing, and rational soil and crop management techniques will all contribute to the improved stability and productivity of the cropping system.

In most areas of the developing world the benefits of the multi-purpose toolbar, above those produced by older conventional equipment, have yet to be proven. In many situations they may not be appropriate, but in others they may be highly relevant and beneficial. Support from governments and external agencies may be necessary to explore the potential for this new equipment fully.

David Gibbon
University of East Anglia

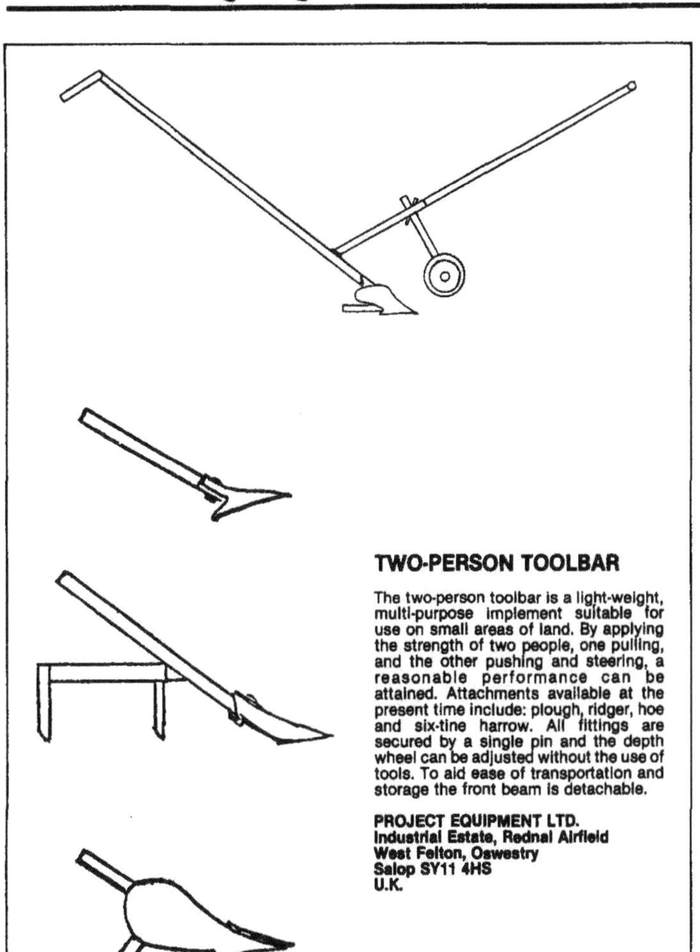

TWO-PERSON TOOLBAR

The two-person toolbar is a light-weight, multi-purpose implement suitable for use on small areas of land. By applying the strength of two people, one pulling, and the other pushing and steering, a reasonable performance can be attained. Attachments available at the present time include: plough, ridger, hoe and six-tine harrow. All fittings are secured by a single pin and the depth wheel can be adjusted without the use of tools. To aid ease of transportation and storage the front beam is detachable.

PROJECT EQUIPMENT LTD.
Industrial Estate, Rednal Airfield
West Felton, Oswestry
Salop SY11 4HS
U.K.

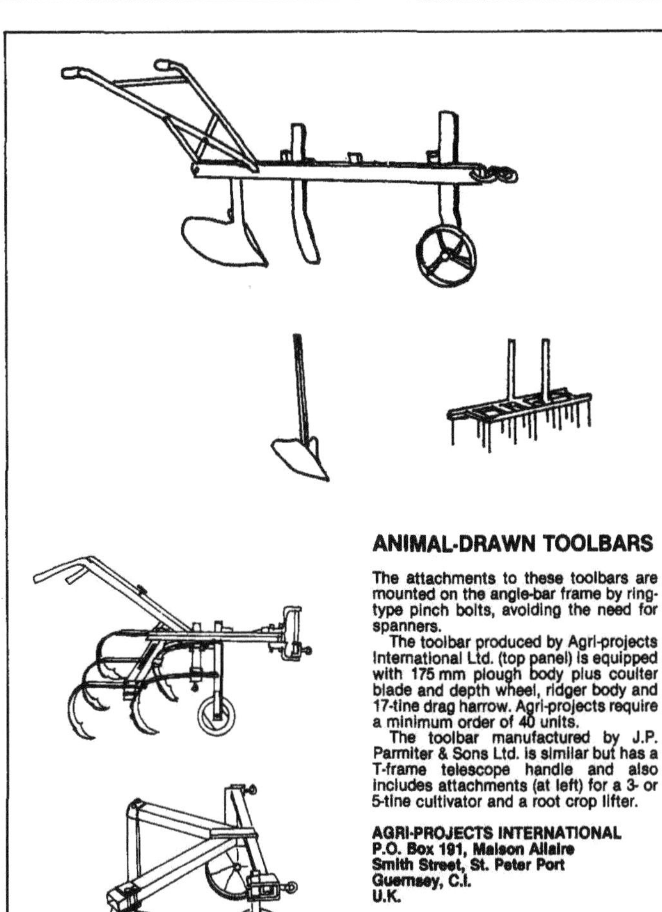

ANIMAL-DRAWN TOOLBARS

The attachments to these toolbars are mounted on the angle-bar frame by ring-type pinch bolts, avoiding the need for spanners.

The toolbar produced by Agri-projects International Ltd. (top panel) is equipped with 175 mm plough body plus coulter blade and depth wheel, ridger body and 17-tine drag harrow. Agri-projects require a minimum order of 40 units.

The toolbar manufactured by J.P. Parmiter & Sons Ltd. is similar but has a T-frame telescope handle and also includes attachments (at left) for a 3- or 5-tine cultivator and a root crop lifter.

AGRI-PROJECTS INTERNATIONAL
P.O. Box 191, Maison Allaire
Smith Street, St. Peter Port
Guernsey, C.I.
U.K.

P.J. PARMITER & SONS LTD.
Station Works, Tisbury
Salisbury, Wiltshire SP3 6QZ
U.K.

MULE-DRAWN TOOLBAR

The equipment, designed by SMECMA for animal draught, is light-weight (12.5 kg) and may be used for ploughing, hoeing/weeding and groundnut lifting. The steel frame and depth wheel give the plough working widths and depths of 10-15 cm and 15-17 cm respectively.

SMECMA
B.P. 1707, Bamako
MALI

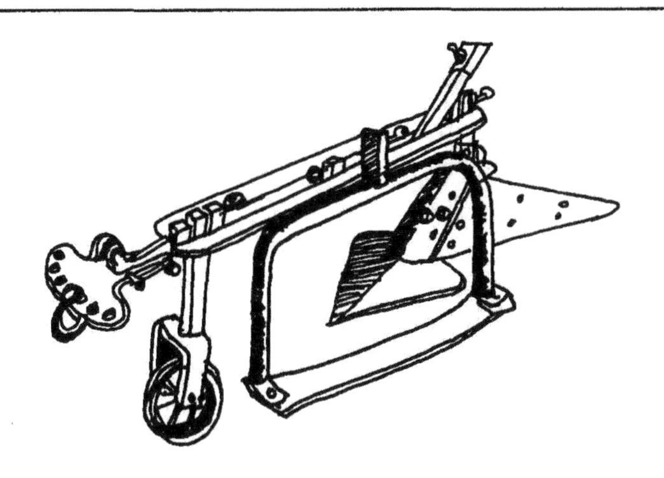

MULTI-PURPOSE TOOL FRAME

An animal-drawn, multi-purpose toolbar comprising a basic frame to which a series of attachments, including plough, ridger, groundnut lifter and cultivator, can be fitted. Further fittings such as soil pulverizers and planters are currently being developed.

AGRI MAL (MALAWI) LTD.
P.O. Box 143, Blantyre
MALAWI

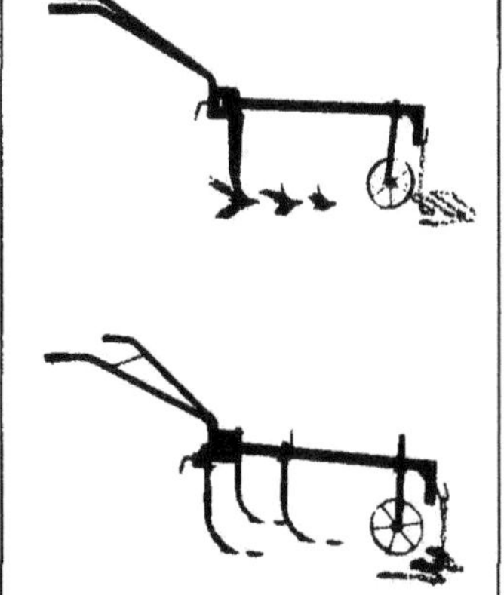

E.B.R.A. OMNICULTOR

E.B.R.A. have produced this simple animal-drawn toolbar to which a wide range of implements may be attached. Shown here are the rigid and spring-tined harrows (top) and the 25 cm mouldboard plough, ridger body and furrower (below). In addition to these the following attachments are available:

1. Digger with 3 different blade widths.
2. Earthing up and ploughing attachment.
3. Two-row planter.

The toolbar frame is constructed of steel, and features adjustable handles and depth wheel.

E.B.R.A., 28 Rue du Maine
B.P. 915, 49009 Angers Cedex
FRANCE

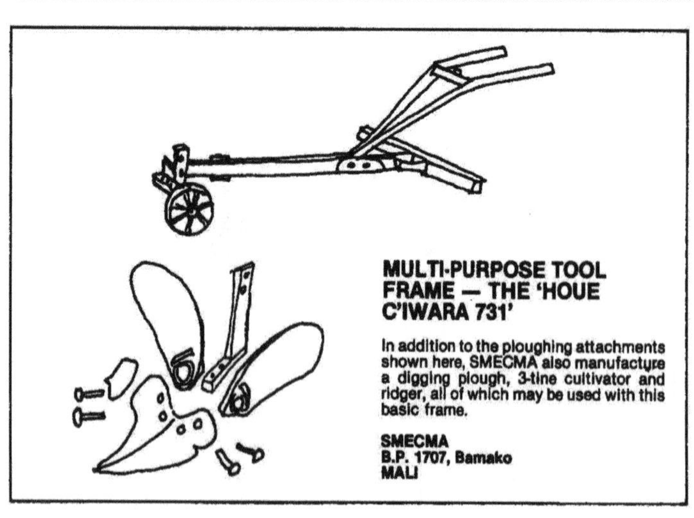

MULTI-PURPOSE TOOL FRAME — THE 'HOUE C'IWARA 731'

In addition to the ploughing attachments shown here, SMECMA also manufacture a digging plough, 3-tine cultivator and ridger, all of which may be used with this basic frame.

SMECMA
B.P. 1707, Bamako
MALI

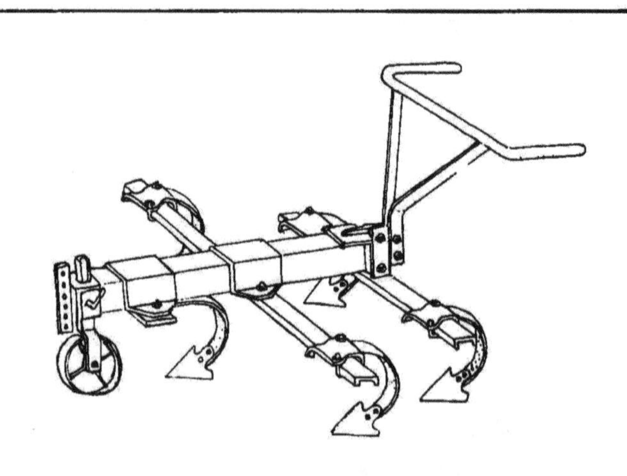

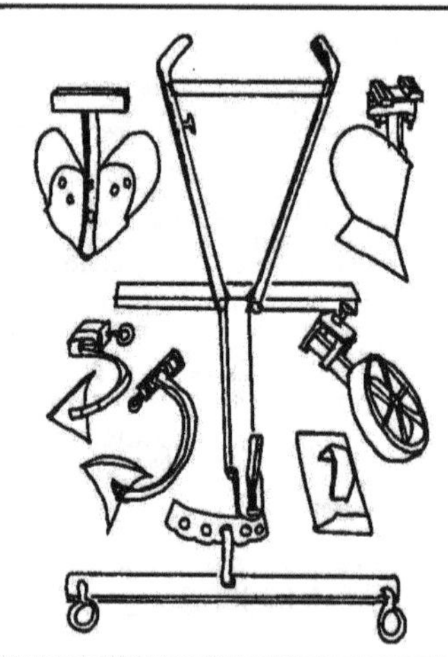

MULTI-PURPOSE, HORSE-DRAWN CULTIVATOR (MPHDC)

The toolbar and attachments developed at the Ministry of Agriculture have recently been introduced into Tongan farming. Production is very limited and so far only five MPHDCs have been incorporated into the demonstration unit of the Extension Division. Project workers are hopeful that the design, based on the tried and tested French 'Houe-Sine' model, will be widely adopted. The MPHDC is a simple, all-steel, construction, equipped with plough, ridger and cultivator. Also featured is an adjustable depth wheel and twin coupling for animal draught. The simplicity of design will allow manufacture and maintenance to be carried out locally.

DEPARTMENT OF AGRICULTURE
P.O. Box 14, Nuku'alofa, TONGA

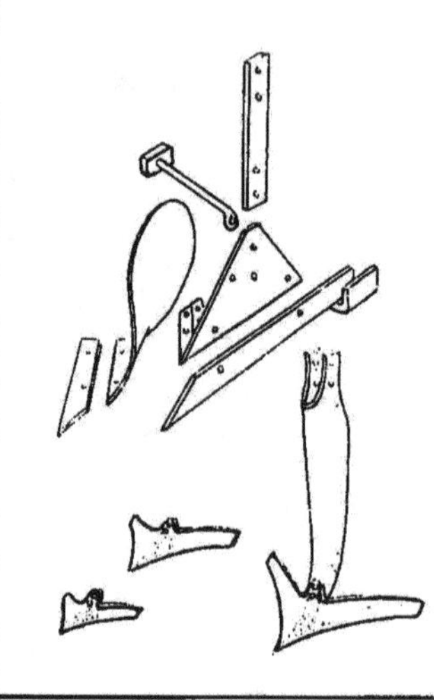

THE 'ARARA' MULTI-PURPOSE FRAME

The 'ARARA' multi-purpose cultivator can be used for different tillage and crop cultivation operations, as well as groundnut lifting. The different attachments are fitted onto the main frame which is equipped with adjustable handles and depth wheel.

The cultivator, including all available attachments, has a total weight of 71.8 kg. This is made up of the following: main frame (19.5 kg); a 25 cm plough-body with draught regulator (16.8 kg); a 3- or 5-tine Canadian cultivator (20.9 kg), ridging plough with adjustable wings and reversible tip (8.8 kg); and a groundnut lifter with 200, 350 and 500 mm blades (5.8 kg).

The above equipment can be ordered in combinations as required.

ARARA
30 Rue d'Anjoy, 78000 Versailles
FRANCE

'SINE 9' MULTI-PURPOSE FRAMES

Seven adaptable units are available: 1 × 8 or 10 HUARD UCF plough-frame, 1 ridging plough with mobile blades, 1 groundnut lifter (3 blades), 1 × 3-tine Canadian hoe or 5-tine with adaptor, 1 excavator pick (3- or 5-teeth), 2 double seed drills, settings from 30 to 90 cm.

Additional features are: vertical setting (for trailing), horizontal setting (for width), equipment coupled with clamps and eye bolts, interchangeable wheel with iron bushing, adjustment and draught chains, weight 30 to 45 kg according to equipment.

Also available is the 'Sine Greco' which uses the same attachments as the 'Sine 9', but has a heavier frame.

SISMAR
B.P. 3214, 20 Rue Dr. Theze, Dakar
SENEGAL

STE NOUVELLE MOUZON
B.P. 26, 60250 Mouy (Oise)
FRANCE

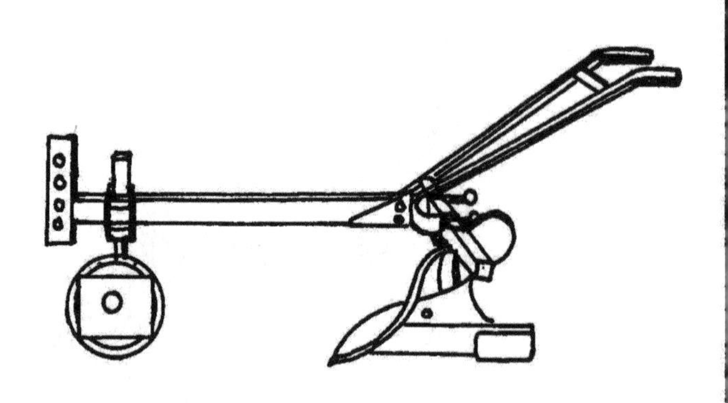

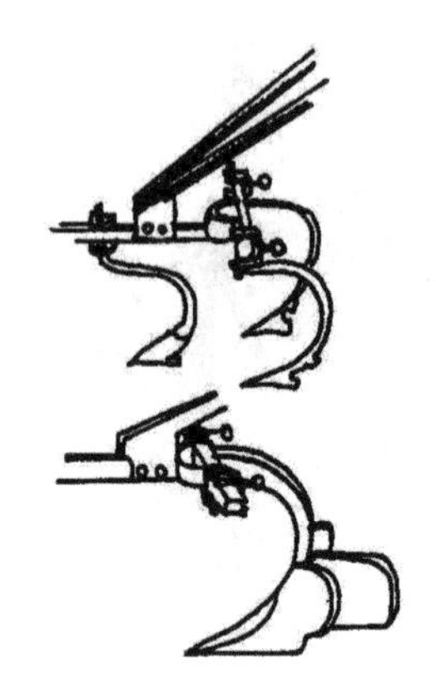

BASIC IMPLEMENT FOR MULTIPLE TILLAGE

The Rau multi-purpose toolbar can be used with a number of attachments for cultivation operations. With reference to the drawings, the components consist of:

1. Basic frame.
2. Gauge wheel.
3. Steel plough body with exchangeable wear parts.
4. Cultivation tines for soil loosening and weed control in row crops.
5. Adjustable ridger for ridge cultivation or furrow-irrigated crops.

There is also a two-row precision seeder (See Section 4).

RAU MASCHINENFABRIK GmbH
Johannes-Rau-Straße
7315-Weilheim an der Teck
W. GERMANY

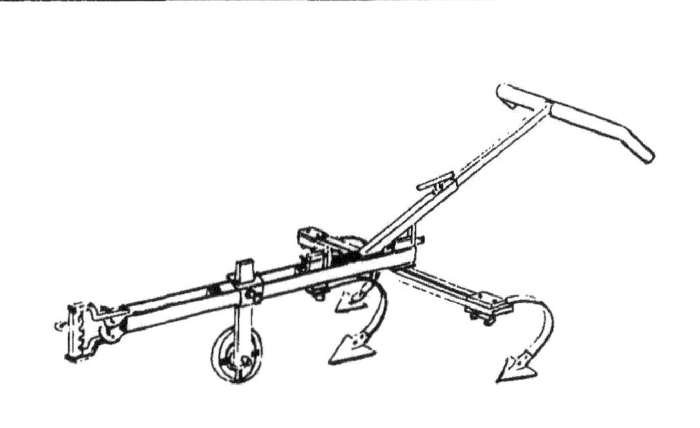

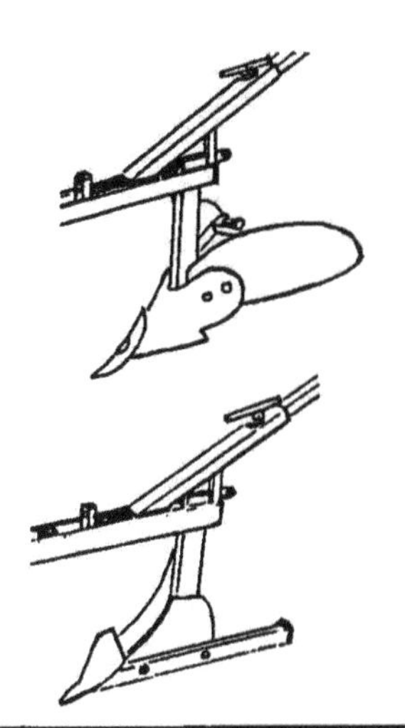

THE ADAPTABAR

A heavy-duty, multi-purpose toolbar strong enough to be used with a 2 or 4 oxen team or pulled by a small tractor. Additional attachments to the 3- or 5-tine cultivator, 22 cm plough and planet ridger pictured here include: roller seeder, groundnut/potato lifter, two-row planter, subsoiler, 5-tine sweep and roller clod breaker. These may be combined where appropriate.

Attachments are retained by ring bolts passing through the stalk, but the thrust is taken by the central frame inserts which are welded in place. Twin A-shaped tube handles can also be supplied.

A heavier frame for hard soil conditions is also available.

PROJECT EQUIPMENT LTD.
Industrial Estate, Rednal Airfield
West Felton, Oswestry
Salop SY11 4HS
U.K.

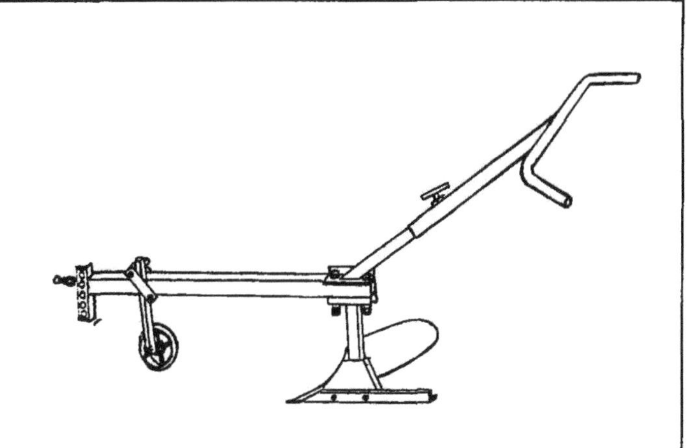

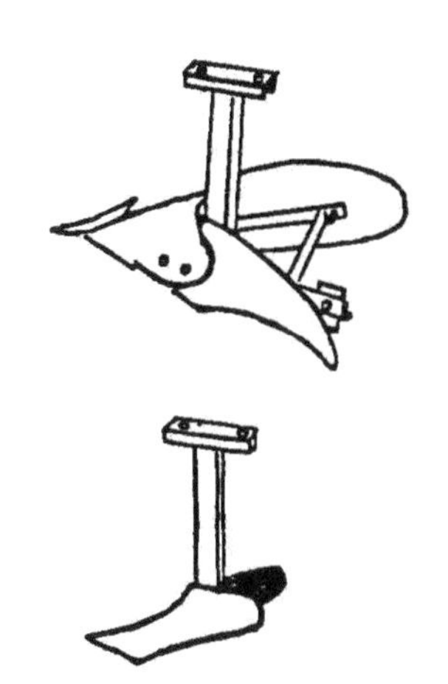

THE PECOTOOL

This is a multi-purpose, animal-drawn toolbar designed specifically for Sierra Leone conditions. The Pecotool was initially produced by Project Equipment, but recently the all-steel implement has been manufactured locally in Sierra Leone itself as part of a Work Oxen Project.

The toolbar can be used with either a 22 cm or 15 cm plough and is supplied with both a wheel and skid. The handle is adjustable, as is the hitch.

Other attachments available are a planet ridger, 3-tine cultivator, groundnut lifter and single row seeder.

PROJECT EQUIPMENT LTD.
Industrial Estate, Rednal Airfield
West Felton, Oswestry
Salop SY11 4HS
U.K.

SIERRA LEONE WORK OXEN PROJECT
P.M.B. 766, Freetown
SIERRA LEONE

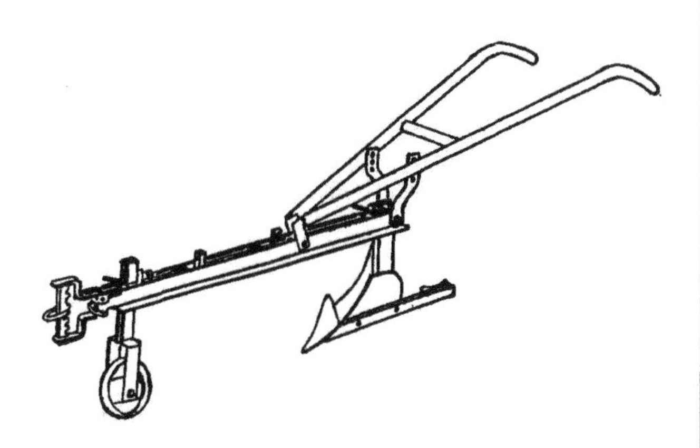

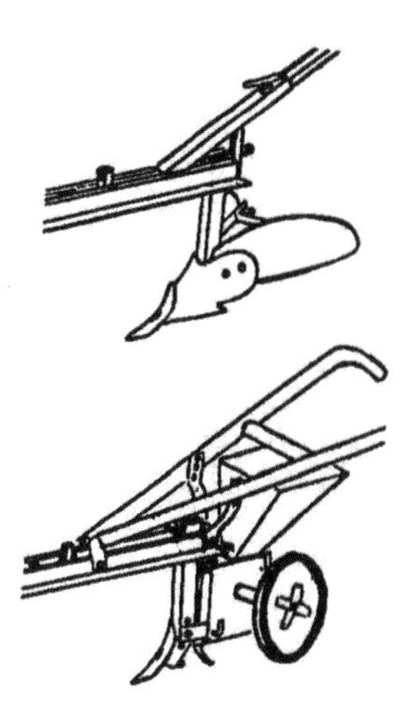

THE ANGLEBAR

Considerably lighter than the Adaptabar, the Anglebar is a multi-purpose model which is within the pulling capacity of smaller draught animals such as ponies, mules, or small oxen.

Available attachments shown are the 22 cm plough, planet ridger and roller seeder.

The Anglebar is manufactured from steel sections which are widely available in many countries, and is intended for production in small workshops possessing only basic equipment.

The main frame is made from angle iron, with handles of water pipe (either A- or T-shaped). The handles are adjustable for height, and the attachments are secured by ring bolts passing through the stalk and mounting points on the frame.

PROJECT EQUIPMENT LTD.
Industrial Estate, Rednal Airfield
West Felton, Oswestry
Salop SY11 4HS
U.K.

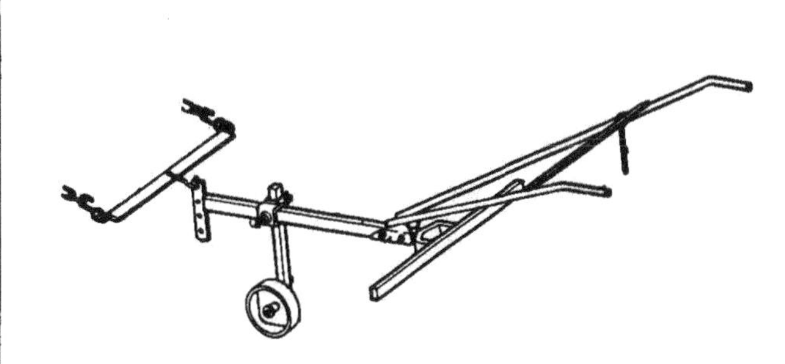

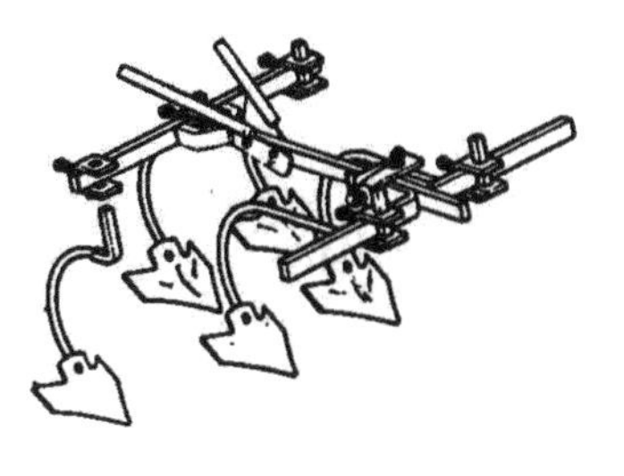

CEMAG POLICULTOR 300

A well-equipped yet simple multi-purpose toolbar, the Policultor 300 can handle up to 7 attachments. Additional to those shown here are a furrower, 5-tine cultivator, drag harrow and seeder.

The Policultor 300 is designed to be drawn by one or two animals and has a basic weight (without attachments) of 24 kg. It is suitable for the cultivation of areas up to 3 hectares.

CEMAG
Rua João Batista de Oliveira 233
06750 — Taboão da Serra
São Paulo
BRAZIL

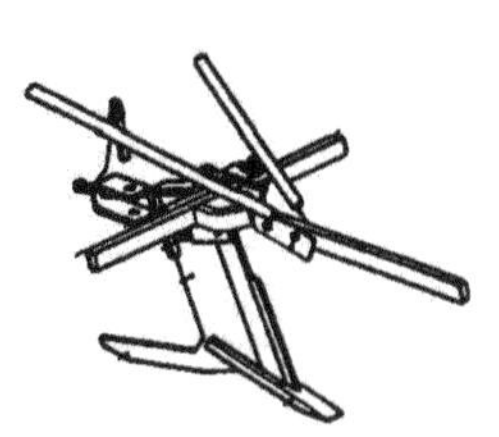

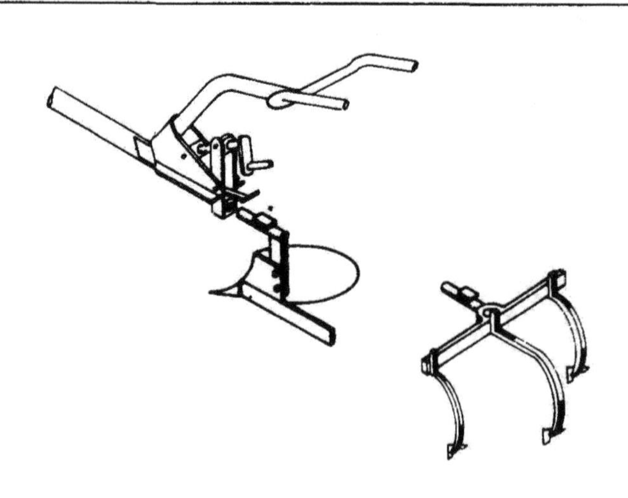

THE KANOL TOOL SHAFT

The Kanol Tool Shaft is available in two forms, one to be drawn by a single animal, the other by two animals. The shafts feature a quick coupling device to which a wide range of attachments may be fitted. These include:

1. Plough
2. 3-tine cultivator
3. Subsoiler
4. Harrow
5. Scraper
6. Furrower
7. Clod crusher
8. Lifter
9. Leveller
10. Seeder

BELIN INTERNATIONAL
2 Mail des Charmilles, B.P. 194
10006 Troyes Cedex
FRANCE

STE NOUVELLE MOUZON
B.P. 26, 60250 Mouy (Oise)
FRANCE

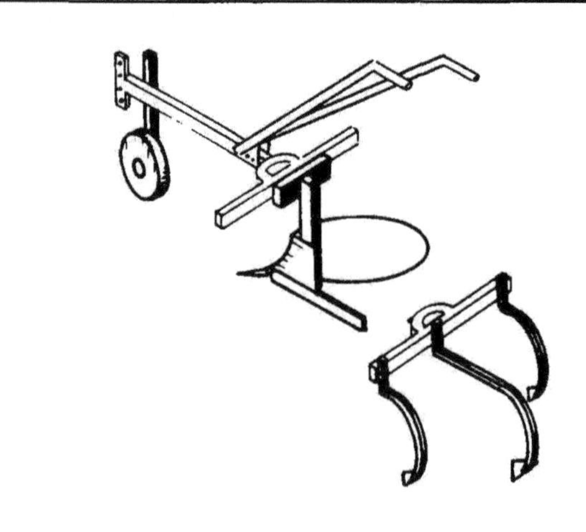

THE SIMONE MULTI-PURPOSE TOOLBAR

The Simone is a simple animal-drawn toolbar with a total of 12 possible attachments. In addition to those shown here, the following are available: tiller/fertilizer-spreader, harrow, clod crusher, seeder, stabilized (2-wheel) plough, and scarifier.

BELIN INTERNATIONAL
2 Mail des Charmilles, B.P. 194
10006 Troyes Cedex
FRANCE

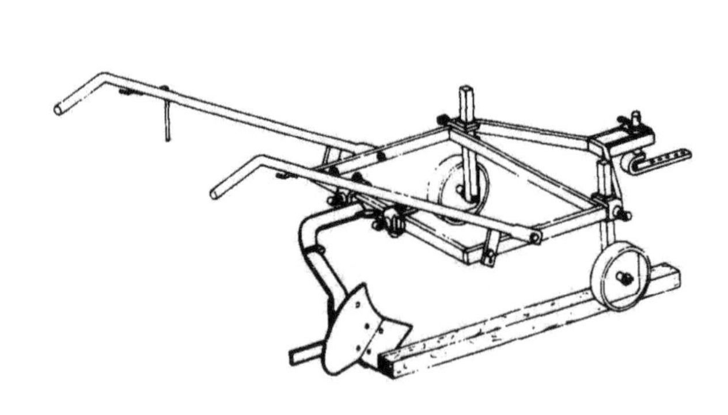

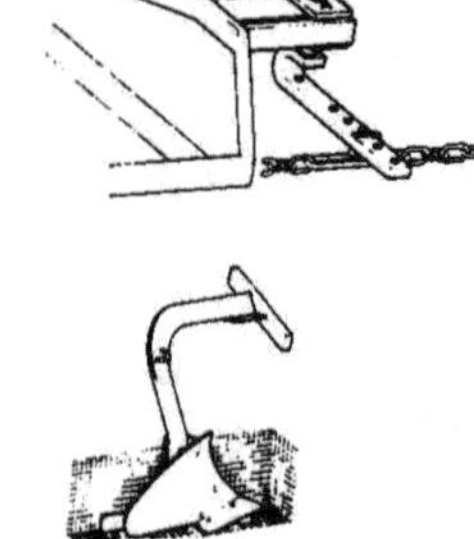

CEMAG POLICULTOR 600

A more advanced model than the Policultor 300, this multi-purpose toolbar is larger and requires greater draught power, and can be fitted with a wider range of attachments. These include

1. Mouldboard plough
2. Furrower
3. Sub-soiler
4. Tiller — with teeth
5. Cultivator
6. Teeth harrow
7. Planter
8. Planter/fertilizer
9. Reversible mouldboard plough
10. Ridger
11. Rake

The Policultor 600 is a 2-wheeled system with a weight (minus attachments) of 48 kg. It is suitable for the cultivation of up to 6 hectare small-holdings.

CEMAG
Rua João Batista de Oliveira 233
06750 — Taboão da Serra
São Paulo
BRAZIL

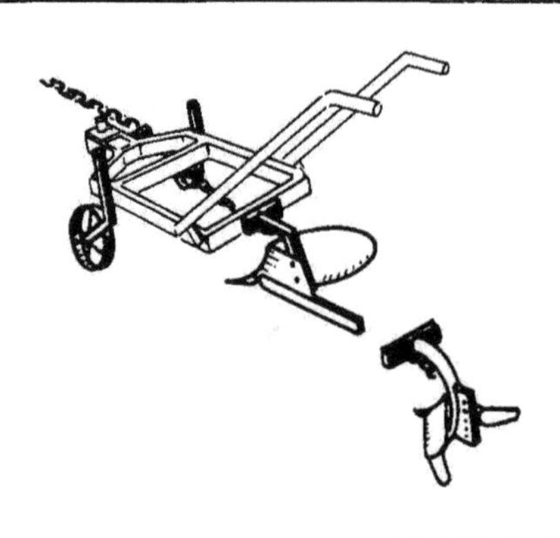

THE MARIANNE MULTI-PURPOSE TOOL BAR

MARIANNE is the third in Belin's range of multi-purpose cultivators, and offers more sophisticated characteristics than the Simone and Kanol. Drawn by 1 or 2 animals, instead of a single-bar frame, the 'Marianne' features a rectangular frame onto which 1 to 3 wheels and one of a choice of 12 implements may be attached. Being larger than the other models it is able to handle double-row attachments and cultivators with a higher capacity than those with 3 tines only.

Available attachments are:

1. Single plough
2. Reverse plough
3. Cultivator (7-tine)
4. Furrower
5. Ridger
6. Peanut lifter
7. Double plough
8. Cultivator (5-tine)
9. Canadian harrow
10. Potato lifter
11. Subsoiler
12. Seeder

BELIN INTERNATIONAL
2 Mail des Charmilles, B.P. 194
10006 Troyes Cedex
FRANCE

THE HOLTAG STRAD

The STRAD animal-drawn, multi-purpose toolbar is designed for ridging, planting, fertilizer application and weeding of crops grown on ridges.

The system consists of the following:
1. Frame.
2. Cross bar.
3. Reversible point and mounting.
4. 2 × 4-gang rotary cultivators.
5. Seed and fertilizer box.
6. Seat and attachment.

JOHN HOLT AGRICULTURAL ENGINEERS LTD.
New Industrial Estate
P.O. Box 352, Zaria
Kaduna State
NIGERIA

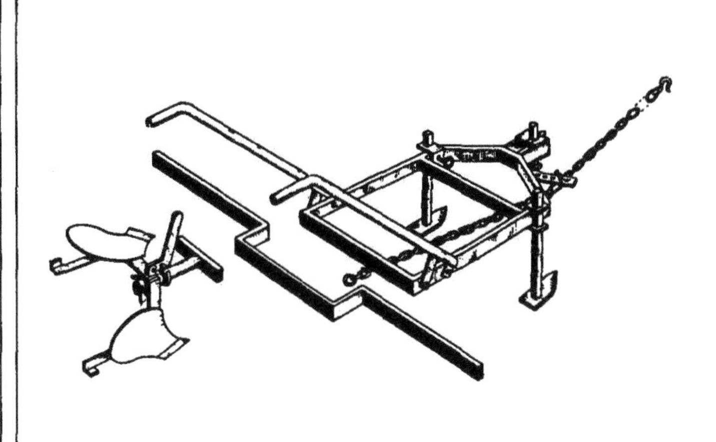

ARIANA MULTI-PURPOSE FRAME

A heavy-duty toolbar, the Ariana is suitable dor difficult soil conditions and rice fields. It features 1 or 2 × 25 cm HUARD UCF cut-away ploughing frames, 1 × 25 cm HUARD UCF quarter turn plough, 1 ridging plough with movable wings, 1 × 3 blade groundnut lifter, 1 × 6- or 8-tine Canadian harrow, 1 × 1.5 m extension bar.

Additional characteristics include: Third wheel at rear of frame, adjustable handles, assembly with clamps and eye bolts, interchangeable wheels with iron bushing. Weight 58-92 kg according to equipment mounted.

The Mouzon Tropiculture also has a disc ridger, a 20-tine harrow, leveller, seeder/fertilizer, sub-soiler etc.

SISMAR
B.P. 3214, 20 Rue Dr. Theze, Dakar
SENEGAL
STE NOUVELLE MOUZON
B.P. 26, 60250 Mouy (Oise), FRANCE

THE BAOL POLYCULTIVATOR

Special features of the animal-drawn Baol Polycultivator include pneumatic wheels, variable guage and swivelling beam. Adaptable units are: 1 × 25 cm HUARD UCF plough frame, 2 ridging ploughs with mobile wings, 2 groundnut lifters (3 blades), 1 × 2 m articulated toolbar, 8 or 12 weeding-hoeing teeth (shown here), 1 × 2 m tip-up cart tray, 1 × 3-row seed drill, Super Eco type.

Additional characteristics: Articulated retractable ploughshare (optional), adjustable handles, large adjustable row-tracer, disconnection and lifting of seed drills, combined ploughshares, space between rows available from 30 to 120cm.

The Baol Polycultivator can also be adapted for use with a small tractor.

SISMAR
B.P. 3214, 20 Rue Dr. Theze, Dakar
SENEGAL
STE NOUVELLE MOUZON
B.P. 26, 60250 Mouy (Oise), FRANCE

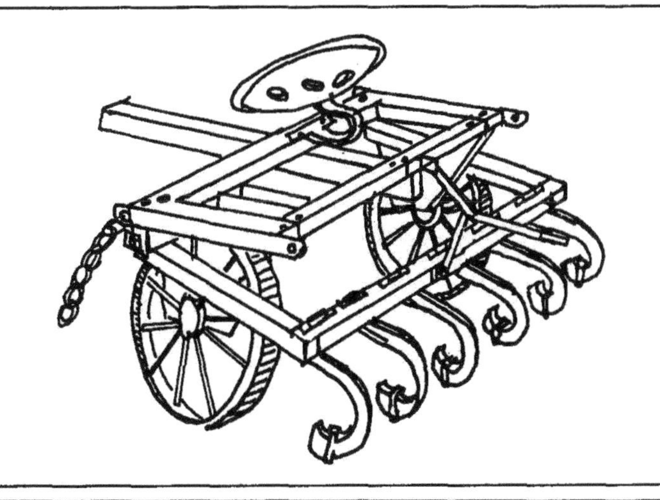

LIONESS ANIMAL DRAWN IMPLEMENTS

The all-steel, fully welded Sahall range of animal-drawn implements can be adapted for ploughing, disc harrowing, cultivating, 2-row ridging or furrowing, 2-row precision seeding, weeding between crop rows, or as an animal-drawn cart with a capacity of up to 500 kg. Pictured left is the main frame fitted with drawbar, operator's seat and toolbar with 6-cultivator tines. A feature of this model is the standardization to metric size M.12 of all bolts and nuts in the mainframe, toolbar and attachments. Only one size of spanner is required to carry out adjustments. In addition, three different types of wheel are available to match the local soil and crop conditions. The wheels are fully adjustable throughout the 3 metre length of the main frame and toolbar.

SAHALL LTD.
Soil and Water Resources
13 Leachfield Industrial Est.
Garstang, Preston PR3 1PR, U.K.

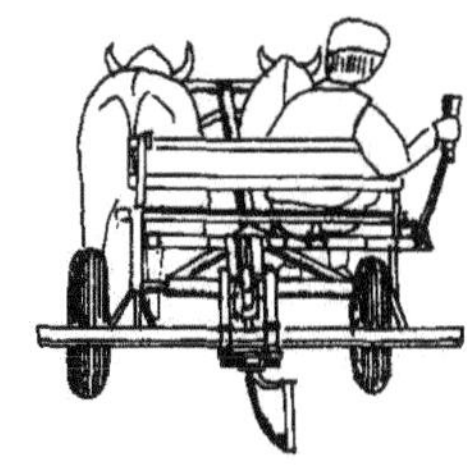

CAPSTA II MULTICULTIVATOR

Developed from the CAPSTA I, this wheeled multi-purpose cultivator may be adapted for use in many field operations, including transportation.

OFICINA VENCEDORA
Av. Sete de Setembro, 599
56.300 Petrolina, PE
BRAZIL

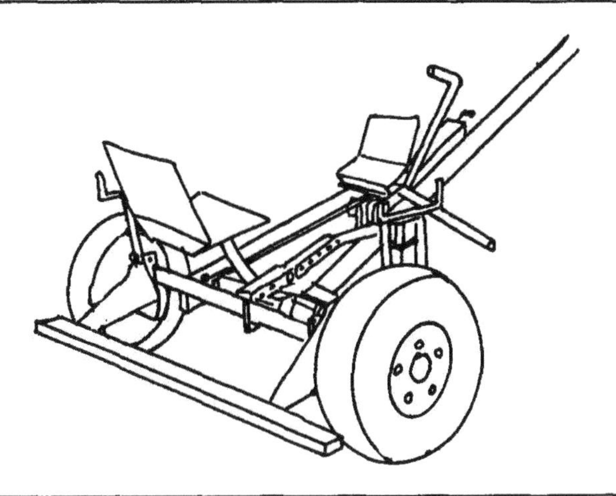

KRUSHI UDYOG MULTI-PURPOSE WHEELED TOOL CARRIER

Suitable for ploughing, cultivation, ridging, inter-cultivation and transportation, the Krushi Udyog Multi-purpose Wheeled Tool Carrier is recommended by the Indian Crop Research Institute for dryland cultivation in deep black vertisole soils. It is drawn by 2 bullocks and comprises the main frame mounted by the operator's seat, two wheels with pneumatic tyres, a mild steel rectangular section toolbar, a spring-loaded adjustment mechanism to control the depth of operation. The multi-purpose tool carrier is most suited to broad bed and furrow cultivation systems.

THE MAHARASHTRA AGRO IND. DEV. CORPN. LTD.
Rajan House, 3rd Floor
Nr. Century Bazaar, Prabhadevi,
Bombay 400 025, INDIA

UNIVERSAL MULTI-PURPOSE TOOL CARRIER

Vicon have designed this animal-drawn tool carrier for a wide range of attachments for land cultivation. These include mould board plough, cultivator, disc harrow, ridger, sub-soiler, and seed and fertilizer applicator.

VICON LTD.
K.R. Puram-Whitefield Road
Mahadevapura Post
Bangalore 560 048, Karnataka
INDIA

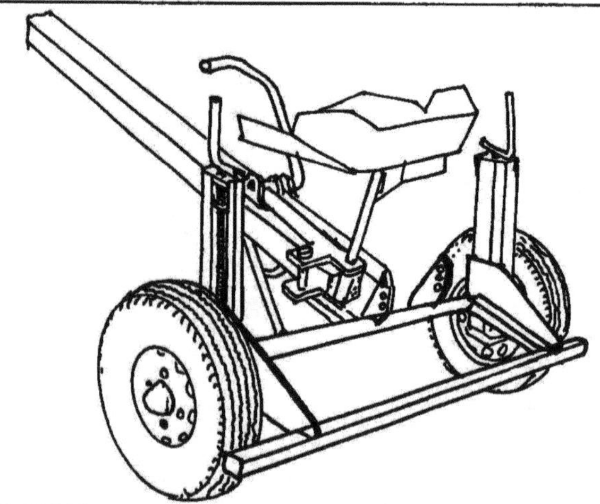

THE NIKART

The Nikart is a wheeled tool carrier consisting of a frame mounted on 2 wheels (usually with pneumatic tyres) with a beam or draw-pole to which a bullock yoke is fastened. The basic frame has a toolbar onto which a variety of implements can be attached with simple clamps. Working depth can be adjusted to meet operational requirements, and a lifting mechanism is provided to raise or lower the implement into position.

The Nikart can be supplied with a range of attachments which enable it to be used for:

1. Tillage:
- Reversible or non-reversible disc or mouldboard plough.
- Ridger.
- Disc, spike or spring-tooth harrow, and cultivator.
- Ridgers and float for bed farming.

2. Planting and fertilizer application:
- Fertilizer application either separately from or in combination with planting.
- Most cereal seeds, peas, beans and cotton can be planted in variable row arrangements and at required spacing. Intercropping can also be carried out.

3. Transport

The Nikart may be obtained from the following manufacturers:

GEEST OVERSEAS MECHANISATION LTD.
Marsh Lane, Boston
Lincolnshire PE21 7RP
U.K.

MEKINS AGRO PRODUCTS (PVT) LTD
6/3/866/A
Begumbed
Greenlands
Hyderabad 500016
INDIA

SRI LAKSHMI ENTERPRISES
65-1 1st Main Road
Ramchandra Purim
Bangalore 560 021
INDIA

VOLTAS LIMITED*
Strand House
19 Graham Road
Ballard Estate
Bombay 400 001
INDIA

SERGIO SOLORZANO DE LA VEGA**
Balboa 125 Esquina Jacarandas
Fraccionamiento Virginia
Veracruz, Ver.
MEXICO

LA VICTORIA S.A.
Ozumba
MEXICO

KALE KRISHI UDYO9
S16/A Narayan Peth,
Pune 411 030
INDIA

GURUNATH INDUSTRIES
65/1 1st Main Road
Ramachandrapuram
Bangalore 560 021
INDIA

CEARA MAQUINAS AGRICOLAS S/A
Av. Gaudioso de Carvalho, 217
Bairro Jardim Iracema
Caixa Postal, D-79
60.000 - Fortaleza - CE
BRAZIL

MEDAK AGRICULTURAL CENTRE (EQUIPMENT)
Cathedral Compound
Medak
Andhra Pradesh 502 110
INDIA

PONTAL MATERIAL RODANTE S/A
Rua Campante No.237
Vila Independencia
Caixa Postal, 833
01.000 - São Paulo - SP
BRAZIL

*Trade name is 'KRISHI RATH'.
**Trade name is 'YUNTICULTOR'.

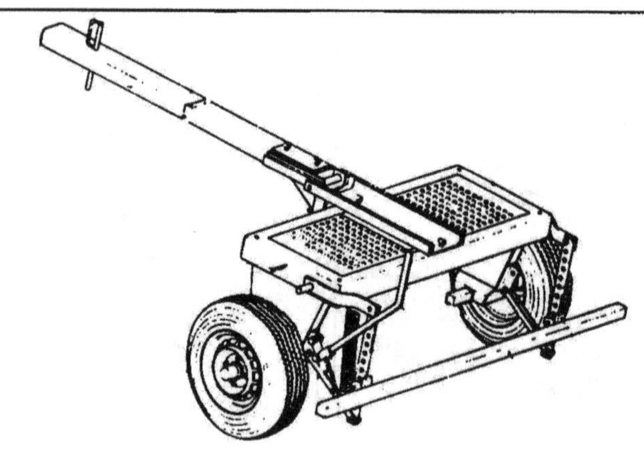

THE AGRIKART

The Agrikart comprises a steel frame mounted on motorcar wheels. A wide range of implements can be attached to the frame, including: mouldboard plough, ridger, cultivator, weeder, disc harrow, seeder, scraper blade, trailer.

These implements are attached to the frame with clamps. Changing implements and adjusting working depth, pitch and wheel track, are possible without the use of spanners. The Agrikart is designed to be drawn by two bullocks and is suitable for use on areas of up to 15 hectares. The Agrikart and similar designs are manufactured by the following companies:

AGRIKART
MEDAK AGRICULTURAL CENTRE (EQUIPMENT)
Medak Agricultural Centre (Equipment)
Cathedral Compound, Medak
Andhra Pradesh 502 110
INDIA

TROPICULTOR
MEKINS AGRO PRODUCTS (PVT) LTD
6/3/866/A
Begumbed
Greenlands
Hyderabad 500016
INDIA

TROPICULTOR
Ste Nouvelle Mouzon
B.P. 26, 60250 Mouy (Oise)
FRANCE

POLICULTOR PONTAL
Pontal Material Rodante S.A.
Rua Campante No. 237
Vila Independencia
Caixa Postal, 833
01.000-São Paulo, SP
BRAZIL

POLICULTOR 1500

This model is equipped with a wider range of attachments which, in addition to those given above, include: furrower, subsoiler, rake, single beam, liquid fertilizer tank, fertilizer spreader, water tank.

CEMAG
Rua Joao Batista de Oliveira 233
06750 — Taboao da Serra, São Paulo
BRAZIL

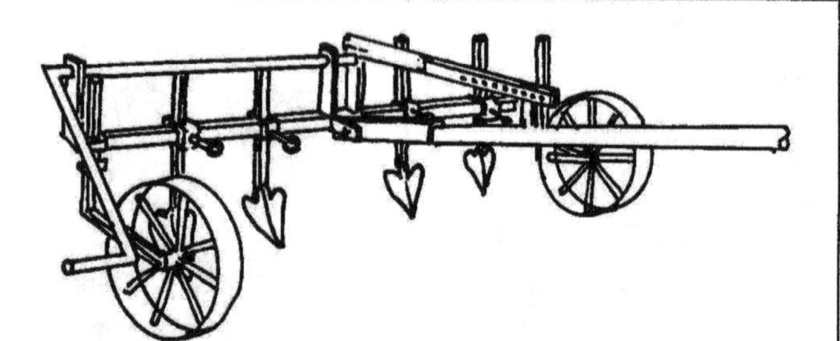

THE AGRIBAR

The Agribar is essentially similar to the Agrikart in that it is equipped with the same 40 × 40 mm square tool bar, and is able to carry the same range of attachments. The Agribar is, however, considerably lighter, and is not provided with pneumatic tyres. These facts allow the Agribar to be cheaper than other toolbars of an equivalent function.

MEDAK AGRICULTURAL CENTRE (EQUIPMENT)
Cathedral Compound, Medak
Andhra Pradesh 502 110
INDIA

POLYNOL ANIMAL-DRAWN AGRICULTURAL TOOL BAR

The Polynol is the most sophisticated of the Belin range of animal-drawn toolbars. For the three chassis types produced for this model, the following attachments are available:

- plough (1 or 2 bodies)
- reversible plough
- cultivator
- harrow
- disc harrow
- sub-soiler
- ridger
- lifter
- ridge leveller
- leveller
- cereal seeder
- fertilizer spreader
- precision seeder
- sprayer
- ridge seeder
- furrow seeder
- grass cutter
- rotary reaper
- charrette
- tip cart
- water tank
- unloading cart
- semi-trailer
- weeder
- hoe
- mixer
- roller
- ridge with disc
- cane-sugar planter
- potato planter

BELIN INTERNATIONAL
2 Mail des Charmilles
B.P. 194, 10006 Troyes Cedex
FRANCE

STE NOUVELLE MOUZON
B.P. 26, 60250 Mouy (Oise)
FRANCE